WHY ME

FROM DREAMS TO DISRUPTION

RIGHT

**UNVEILING THE
EXTRAORDINARY JOURNEYS OF**

NOW?

**7 LATINTECH START-UP
ENTREPRENEURS**

JAVIER CUELLO   CHRISTY FERNANDEZ-CULL
FERNANDO JOHANN   NICK LOPEZ   DANIEL PARDO
HANS VILLA   ISMAEL VILLANUEVA

FOREWORD MARIELA ROMERO

Advance Praise

When Tampa Bay Wave and the Tampa Bay Latin Chamber leaders approached me with their vision to empower Hispanic entrepreneurs who are underrepresented in the tech sector, I jumped at the opportunity to champion seed funding to make their vision a reality and bring higher-paying jobs to our region.

Tampa Bay Wave's LatinTech Accelerator is blazing pathways of opportunity for innovators—and doing it in an inclusive way. We aim to boost tech startups led by Hispanic entrepreneurs and add to the Tampa Bay area's tech startup boom. Together, we are going to grow businesses, solve problems and ensure that the Tampa Bay economy continues to thrive.

> Kathy Castor
> US Representative (FL14)

We are excited to see this Tampa Bay Wave Accelerator advance Tampa's long history of Latin-heritage entrepreneurs. This book hopefully not only brings recognition to these entrepreneurs but inspires others to reach for their dreams.

> Julio Esquivel
> Attorney at Shumaker, Loop & Kendrick, LLP

On behalf of Visit Tampa Bay and myself, I am delighted to recognize the Tampa Bay Wave Accelerator as a leading-edge producer of innovative and entrepreneurial startups in the Tampa Bay area. The strides Wave has made in supporting Latin-owned businesses to create bridges and connect Tampa Bay's rich Latin heritage to the expansive future of our community are truly inspirational.

Santiago C. Corrada
President and CEO, Visit Tampa Bay

Our collaboration with Tampa Bay Wave to launch the LatinTech Accelerator is a testament to our innovative approach. This program is designed to attract leading startups from Latin America and our local region, creating a melting pot of innovation right here in Tampa. The dream was big.

Cesar R. Hernandez
CEO, Founder and Managing Director, Omni Public
Co-Founder, LatinTech Accelerator
Chairman, Tampa Bay Latin Chamber of Commerce

The LatinTech Accelerator, in partnership with the Tampa Bay Latin Chamber and with vital support from US Representative Kathy Castor and the US SBA, marks a transformative step for tech diversity. We celebrate the ingenuity of our founders of Latin heritage, whose stories illuminate our mission. This program is more than an initiative; it's a commitment to inclusivity, innovation, and community empowerment in the tech industry.

Linda Olson
President and CEO, Tampa Bay Wave

Copyright © 2024 by NOW SC Press

All rights reserved. No part of this publication may be reproduced, distributed, or transmitted in any form or by any means, including photocopying, recording, or other electronic or mechanical methods, without the prior written permission of the publisher, except in the case of brief quotations embodied in critical reviews and certain other noncommercial uses permitted by copyright law. For permission requests, write to the publisher, addressed "Attention: Permissions Coordinator," via the website below.

Publish@nowscpress.com
www.PublishWithNOW.com
@nowscpress

Ordering Information:

>Quantity sales. Special discounts are available on quantity purchases by corporations, associations, and others. For details, contact the publisher at the address above.

>Orders by U.S. trade bookstores and wholesalers. Please contact: NOW SC Press: Tel: (813) 970-8470 or visit www.PublishWithNOW.com

Printed in the United States of America

First Printing 2024

Dedication

This book is dedicated to Manny T. Garcia, whose first corporate career was as an AT&T technician. His various life experiences have shaped his profound appreciation for formal education.

As a trailblazer in his family, he became one of the first college graduates, setting a powerful example for future generations.

Manny's exceptional corporate leadership skills, honed through the years earned him immense respect within the Fortune 100 companies he worked for. His remarkable achievements as a true MBA have left an indelible mark on those fortunate enough to have worked alongside him. We should all be so fortunate to have a mentor and business advisor such as Manny T. Garcia.

Contents

Preface 1
    Cesar R. Hernandez, Co-Founder, LatinTech Accelerator

Introduction 9
    Dr. Richard Munassi, Managing Director, Accelerator Program, Tampa Bay Wave

Foreword 13
    Mariela Romero, Emmy® Award-winning Journalist, Author, International Speaker and Co-Founder, Conferencia de Transformación Latinoamericana

Chapter 1 19
    TELL's Critical Mission
    Fernando Johann, Founder, TELL

Chapter 2 29
    A Proud Badge of Recognition
    Ismael Villanueva, Founder, Miss Berni

Chapter 3 39
Pioneering Everyday Health Monitoring with Wearables
Christy Fernandez-Cull, Founder and CEO, Davinci Wearables

Chapter 4 49
Revolutionizing Recruitment
Daniel Pardo, Co-Founder, Hitch

Chapter 5 59
The Potential of Latino Talent in Tech
Nick Lopez, Founder, Prosal

Chapter 6 69
Nufi's Important Work
Hans Villa, Co-Founder and CEO, Nufi

Chapter 7 79
Global Growth and Learning from Experience
Javier Cuello, Co-Founder and CEO, H+Trace

Words Change Bias 89

What's Your Number: Epilogue 91
Liza Marie Garcia

NOW Publishing & Publicity 103

Tampa Bay Latin Chamber of Commerce 105

Tampa Bay Wave 107

Preface

When I decided to start the Tampa Bay Latin Chamber with Roberto Borrero, it was from a place of genuine need we saw in our community. Roberto, with his experience from Mobile Apps Media, and I shared a vision to support Latino entrepreneurs, especially those with a spark for tech. The idea wasn't just to fill a gap; it felt more like bridging a wide chasm where Latino tech founders were often left to navigate the waters alone.

The dream was big—a LatinTech Accelerator that could not only support our local talent but also attract innovative minds from all over Latin America. But *dreams have a way of showing you the reality of your own limitations.* As much as we wanted the

Chamber to spearhead this, we knew deep down we weren't set up for something as specialized as a tech accelerator.

That's when we turned to Linda Olson and Dr. Richard Munassi. Their backgrounds made them the perfect allies, and thankfully, they saw the potential in our idea. The thought of Tampa Bay Wave housing our accelerator brought a new surge of hope.

Yet there was still a mountain of funding to climb. That's where Congresswoman Kathy Castor came in, Her support, far from mere luck, felt like destiny unfolding. Her commitment and backing of $500,000 through her community funding program was not just a financial boost but a profound affirmation of our collective vision. Her belief in our project was the final piece of the puzzle. It wasn't just about the money. It *was her vote of confidence that really made the difference.* It made our ambitious project feel possible as well as personal, rooted in the community's belief in us and in the potential of Latino entrepreneurs here in the Tampa Bay area.

As Tampa evolves, destined to become the capital of Latin tech, future generations will look back and recognize this moment as the **spark that ignited a transformative era**. Our efforts today, grounded in community and driven by a shared vision, will be remembered as the beginning of an extraordinary chapter in our history.

The inception of the LatinTech Accelerator within the Tampa Bay Latin Chamber marks a significant milestone for several reasons, signaling a transformative moment for the local Latino community and also for the broader tech ecosystem in the Tampa Bay area and beyond.

First, this initiative represents a pioneering effort specifically tailored to uplift Latino entrepreneurs in the tech sector. By acknowledging and addressing the unique challenges faced by Latino tech founders, the program sets a precedent for targeted support that goes beyond general entrepreneurship aid.

It recognizes the cultural, linguistic and systemic barriers that Latino entrepreneurs often encounter and provides a tailored framework to navigate and surmount these hurdles.

Second, the LatinTech Accelerator is significant for its role in fostering diversity and inclusion within the tech industry, a sector that has historically struggled with representation. By empowering Latino founders, the program enriches the tech landscape with a diversity of ideas, perspectives and innovations that are essential for driving forward-thinking solutions and *fostering a more inclusive and equitable tech community.*

Moreover, the collaborative nature of this initiative—bringing together the vision and resources of the Tampa Bay Latin Chamber and Tampa Bay Wave, and the support of Congresswoman Kathy Castor—highlights the power of community and cross-sector partnerships. This synergy not only amplifies the impact of the program but also serves as a model for how collaboration can bridge gaps between different sectors and communities.

The LatinTech Accelerator is a beacon of opportunity for the local Latino community and for aspiring tech entrepreneurs from Latin America and beyond. It positions Tampa as a nurturing hub for Latino innovation, with the potential to attract talent and ideas from across the globe, thereby enriching the local economy and tech ecosystem.

WhyMeRightNOW.com

In essence, the launch of this program is a foundational step toward a more inclusive, diverse and vibrant future for the industry. It heralds *a new era in which the contributions and potential of Latino tech entrepreneurs are recognized, nurtured and celebrated*, paving the way for a more inclusive and dynamic tech landscape in the years to come.

Embarking on this journey to bring the LatinTech Accelerator to life, I reached out to numerous Hispanic/Latino support organizations, hoping to find allies who shared our vision and enthusiasm. These organizations, with their deep roots and commendable missions, seemed like natural partners for an initiative aimed at empowering Latino tech entrepreneurs. Yet, as we presented our blueprint for change, a stark realization dawned upon us. The concept of a tech accelerator, specifically tailored for the Latino community, was not just novel but entirely foreign to their traditional frameworks. Their hesitance wasn't rooted in a lack of goodwill but in adherence to established paths that had yet to venture into the uncharted territories of tech innovation and entrepreneurship.

This response, while initially disheartening, became a clarion call for us. It underscored a gap not just in the availability of resources but also in the collective imagination of what could be achieved for and by the Latino community within the tech sphere. The reluctance of these organizations to venture beyond their comfort zones highlighted a broader issue: a systemic underestimation of the transformative potential of Latino entrepreneurs in the tech industry.

Thus, we found ourselves at a crossroads, faced with the choice to conform to the existing landscape or to forge a new path. The decision was clear. We chose to set a new precedent, to step into the void left by traditional support structures and *create something that was not just needed but was revolutionary*. Establishing and funding the LatinTech Accelerator outside the conventional Hispanic support networks was not an act of defiance but a bold statement of belief in the untapped potential of Latino tech founders.

This decision to move forward independently was profound, marking a significant shift in how support and empowerment for Latino entrepreneurs were conceptualized. It was a commitment to not only fill an existing gap but also to expand the horizon for what our community could aspire to and achieve. In doing so, we laid the groundwork for a new ecosystem, one where Latino tech entrepreneurs are participants as well as leaders and innovators, shaping the future of technology for their community—and the world.

We encountered other obstacles that propelled us in our determination to redefine the narrative around Latino tech entrepreneurship. Understanding the task at hand was monumental: to create from the ground up an ecosystem that acknowledged and actively championed the unique contributions and needs of Latino tech founders. This meant building not just a program but a community and a network from scratch, without the immediate backing of established entities that many might consider essential for such an endeavor.

Securing funding in this uncharted territory was another significant challenge. Without the immediate support of traditional Hispanic/Latino organizations, we had to cast a wider net, engaging with a diverse range of stakeholders who shared our vision for a more inclusive tech landscape. *This journey was marked by countless pitches, meetings and*

WhyMeRightNOW.com

advocacy efforts to convey the importance and potential impact of the LatinTech Accelerator.

Moreover, the endeavor to create a space where Latino tech entrepreneurs could thrive was not just about logistical and financial hurdles. It was also about challenging and changing deep-seated perceptions within and outside the Latino community—about what we can achieve in the tech industry and how we support one another in reaching those heights.

Despite these obstacles, or perhaps because of them, the journey to establish the LatinTech Accelerator has been profoundly transformative. Each challenge we faced underscored the critical need for such an initiative and fueled our commitment to pave a new path for Latino entrepreneurs in tech. Through perseverance, collaboration and an unwavering belief in our mission, we've begun to turn the tide, laying the foundation for a future where Latino tech founders are recognized as integral and innovative leaders within the global tech community.

In the face of the challenges encountered in establishing the LatinTech Accelerator, the journey has been a testament to resilience, innovation and the enduring spirit of the Latino community. The initial skepticism from traditional support organizations and the daunting task of securing funding did not deter our mission; rather, **they illuminated the path with a clarity** that our endeavor was not just necessary but vital. Each obstacle became a stepping stone, each setback a lesson in perseverance, driving us forward with an unwavering commitment to carve out a space where Latino tech entrepreneurs could flourish.

This journey, marked by both its trials and triumphs, continues a legacy that runs deep in the roots of Latino history—a legacy of innovation, ingenuity and immense contributions to society. Our ancestors, from the advanced civilizations of the Aztecs with their intricate calendars and sprawling cities, to the Maya with their profound understanding of mathematics and the cosmos, to the Incas with their extensive network of roads surpassing even the greatest Roman achievements, have laid the foundations of a rich heritage of making monumental contributions to global society.

Drawing inspiration from this remarkable heritage, the LatinTech Accelerator stands as a beacon of what the new generation of Latino entrepreneurs is capable of contributing to the global society, particularly in the realm of technology. It serves as a reminder that the spirit of innovation, resilience and community that guided our ancestors continues to pulse through the veins of today's Latino entrepreneurs.

As we look to the future, the LatinTech Accelerator is more than a program. It is a movement symbolizing a new chapter in which the Latino community continues to shape the world with groundbreaking innovations in tech. It is a call to action for the new generation to embrace their heritage, to build upon the legacy of their ancestors and to make their mark on the world. The challenges we've faced in bringing this vision to life are dwarfed by the potential impact of empowering Latino voices in technology, inspiring a wave of innovation that will resonate for generations to come.

In this light, the story of the LatinTech Accelerator is one of hope, inspiration and an unwavering belief in the potential of the Latino community to contribute to and transform the global

WhyMeRightNOW.com

tech landscape. It is a tribute to our ancestors, a commitment to the present and a gift to the future, urging every Latino entrepreneur to dream big, persevere and continue the legacy of making significant contributions to society.

Cesar R. Hernandez
CEO, Founder and Managing Director, Omni Public
Co-Founder, LatinTech Accelerator
Chairman, Tampa Bay Latin Chamber of Commerce

Introduction

Welcome to *Why Me, Right NOW?*, a narrative that encapsulates the entrepreneurial spirit of seven founders who are at the forefront of reshaping the tech industry. This book is not just a collection of stories; it's a reflection of a transformative movement from Tampa Bay Wave in collaboration with the Tampa Bay Latin Chamber of Commerce. Through the LatinTech Accelerator program, we've set out to leverage the underrepresented potential within the Latino community worldwide, aiming to foster an environment where innovation and heritage intersect to create something truly groundbreaking.

The inception of the LatinTech Accelerator was marked by the recognition of a significant imbalance in the representation of founders of Latin heritage in the tech sector. Despite constituting a substantial and rapidly growing demographic within the US, their presence in the tech landscape and STEM fields remains disproportionately low. With the support of a $500,000 federal grant, championed by US Representative Kathy Castor and the US Small Business Administration, the LatinTech Accelerator was established in 2023 to help change those statistics. This initiative was not just about providing resources; it was about nurturing a culture of innovation and inclusivity.

The program stands out for its comprehensive approach to fostering growth and excellence. Participating companies benefit from an accelerator curriculum meticulously designed to address the unique needs and aspirations of entrepreneurs of Latin heritage. From executive coaching and mentorship from industry leaders to strategic networking opportunities and a platform to showcase their innovations, the LatinTech Accelerator is a testament to Tampa Bay Wave's commitment to cultivating a thriving, inclusive tech ecosystem.

This book marks a significant milestone, not just for the individuals it features but also for the broader community it represents. The stories of these seven entrepreneurs are a powerful reminder of the creativity, resilience and vision that characterize the community. Their journeys, though diverse in their paths and strategies, converge on a shared destination of innovation, growth and success.

As you immerse yourself in the pages of *Why Me, Right NOW?*, you are invited to witness the unfolding of an evolving chapter in the tech industry—one where diversity is not just acknowledged but celebrated. These narratives are more than success stories; they are blueprints for aspiring entrepreneurs and visionaries worldwide. They underscore the potential that lies in embracing one's heritage, persisting through uncertainties and relentlessly pursuing one's vision.

In a world where the landscape of technology and entrepreneurship is ever evolving, *Why Me, Right NOW?* stands as a testament to the power of representation, the importance of community support and the boundless potential of dreams backed by action and determination. This book is an invitation to be inspired, to be motivated and to be a part of the extraordinary journey of these LatinTech entrepreneurs. Welcome to a celebration of innovation, heritage and the indomitable spirit of the Latino community. Welcome to *Why Me, Right NOW?*

Dr. Richard Munassi
Managing Director, Accelerator Program,
Tampa Bay Wave

Foreword

The reverberations of the COVID-19 pandemic still resonate within me, not merely as echoes of a global crisis but as a clarion call that awoke a dormant realization in my soul. To me, as a seasoned journalist and a proud member of the vibrant Latino community in the United States of America, the pandemic was more than a news story; it was a personal encounter with vulnerability and untapped potential. Witnessing the disproportionate impact on my community, I felt a profound need to transform my role from an observer to an active catalyst for change.

For over three decades, my career in broadcasting has been a journey of connecting stories and people, a bridge between events and audiences. Yet a new calling emerged as

I navigated the pandemic's aftermath. I yearned to transition from the role of a storyteller to an enabler, from informing to empowering. This marked the genesis of a new chapter in my life—a shift toward creating tangible platforms for growth and transformation for the Latino community.

Embracing this new path meant leaving behind the comfort and predictability of a successful broadcasting career. It was a leap into the unknown, driven by a vision to create a significant impact. This journey led to authorship, speaking and, most significantly, to the birth of the *Conferencia de Transformación Latinoamericana* (Tampa and Houston in 2022, Atlanta 2023–2024)—a convergence point for ideas, innovation and empowerment. The conference evolved into more than a gathering; it became a crucible of transformation, blending the resilience and ingenuity of the Latino spirit with practical tools for entrepreneurship and leadership. The conference also taught me, as an entrepreneur, how this journey is often romanticized, yet in reality the path is fraught with challenges.

In this foreword, I am honored to introduce *Why Me, Right NOW? From Dreams to Disruption: Unveiling the Extraordinary Journeys of 7 Latin Tech Startup Entrepreneurs,* a book that transcends the confines of any conference or event. It is a deep dive into the lives of extraordinary Latino tech entrepreneurs. Their stories are not just narratives of success; they are blueprints for navigating the complexities of the startup world, enriched by the resilience, innovation and cultural richness that define the Latino community.

Each chapter of this book unfolds a unique journey, revealing the trials and triumphs of these founders. Their experiences offer a candid look into the startup world, an often underrepresented yet profoundly inspiring narrative. These stories are a testament to the enduring spirit of Latino entrepreneurship, demonstrating how cultural richness and adversity can be powerful catalysts for innovation and success.

Through *Why Me, Right NOW?* readers are invited to step into the lives of these seven entrepreneurs to share in their moments of vulnerability, breakthroughs and epiphanies. The book serves as an intimate exploration of what it truly takes to build a tech startup from the ground up, presenting lessons in leadership, innovation and the power of diversity.

Throughout my life's journey, spanning continents from my birthplace in Venezuela to France, Mexico, Argentina to eventually finding a permanent home in the Tampa Bay area as a citizen of the United States, I have been a perpetual student of culture. This rich tapestry of experiences has profoundly shaped my understanding of the world and the people who inhabit it. What struck me most profoundly in the narratives of these seven entrepreneurs within *Why Me, Right NOW?* is the vivid imprint of their cultural heritage on their entrepreneurial paths. Each founder brings to their venture a unique cultural richness—a fusion of values, norms, language, symbols and beliefs that are intricately woven into the very fabric of their business endeavors.

These cultural influences are not just passive backdrops; they are active, dynamic forces that shape decisions, drive innovation and inspire solutions. As I delved into each story, I observed how these entrepreneurs drew strength from their cultural roots, channeling the essence of their identities into the creation and evolution of their products and services. Their cultural lens offered a distinctive perspective, transforming

WhyMeRightNOW.com

perceived obstacles into opportunities for growth and innovation. It was as if each entrepreneur's background served as a compass, guiding their journey through the complex terrain of the tech startup world. The richness of their cultural experiences added depth and color to their business strategies, enabling them to navigate challenges with a unique blend of resilience and creativity.

This cultural dimension goes beyond mere influence; it is a testament to the power of diversity in the entrepreneurial landscape. It highlights how the amalgamation of different cultural experiences can lead to groundbreaking ideas and approaches, setting these entrepreneurs apart in a highly competitive market. Their stories underscore the reality that cultural diversity is not just a buzzword but a tangible asset that enriches every facet of business, from product development to market engagement. In a world that is increasingly interconnected, the ability to draw upon a diverse cultural heritage is more than an advantage—it is a necessity for success and innovation.

As I reflect on these narratives, I am reminded of my own cultural journey and how it has shaped my understanding of the world. The stories in *Why Me, Right NOW?* resonate with a deeper truth—that our cultural experiences are not just part of who we are; they are powerful tools that can drive change, foster innovation and create a lasting impact. This book is not just a celebration of entrepreneurial success; it is a vibrant mosaic of cultural richness, a showcase of how the diverse tapestry of human experiences can come together to create something truly extraordinary. This book is a demand to recognize and harness the potential within the Latino community. It is a narrative that redefines the contours of success, urging readers to dream, achieve and inspire. Each story within these pages is a beacon of hope, shining a light on the path to empowerment and transformation.

Why Me, Right NOW? challenges conventions and redefines what it means to be a disruptor in the tech industry. It is an affirmation of the entrepreneurial spirit that thrives within the Latino community, serving as an inspiration for current and future generations of dreamers and doers.

As you embark on this journey through the stories that follow, prepare to be inspired, moved and transformed. This book is not just a collection of individual successes but a collective narrative of resilience, innovation and the indomitable spirit of Latino entrepreneurs.

Welcome to a transformative experience that celebrates the Latin American spirit and offers a blueprint for the future. Together, let us explore the extraordinary paths of these seven entrepreneurs, learning from their experiences and drawing inspiration for our own journeys. This foreword is just the beginning of an exploration into the power of diversity, the resilience of the human spirit and the endless possibilities that await when we dare to ask, "Why not me?"

Mariela Romero
Emmy® Award-winning Journalist, Author,
International Speaker and Co-Founder,
Conferencia de Transformación Latinoamericana

WhyMeRightNOW.com

CHAPTER 1

TELL's Critical Mission

> " It's about ensuring all people can understand their brain health. "
> —Fernando Johann, Founder, TELL app

Never before has the need for TELL's (Toolkit to Examine Lifelike Language) groundbreaking services been so pressing, and here's the compelling reason why. At the forefront of our concerns, the escalating urgency to unearth innovative solutions for neurodegenerative diseases is paramount. These afflictions are on the rise, particularly in under-researched regions such as Latin America, making our mission more critical than ever.

TELL's unique approach, leveraging artificial intelligence for speech analysis, aligns perfectly with the increasing emphasis on personalized and accessible healthcare solutions. The convergence of advancements in technology, the pressing need for early and accurate diagnosis and the collaborative efforts of international research initiatives, such as the one led by TELL's Adolfo García, all contribute to creating a conducive environment for TELL's services.

Noninvasive, language-based assessment for brain health

TELL's capability to offer a noninvasive, language-based assessment for brain health is a significant key to make brain health affordable and accessible. The current landscape recognizes the importance of such nonintrusive methodologies in medicine, especially when considering diverse patient populations and the need for scalable, cost-effective solutions in and outside Latin America.

The confluence of technological readiness, the global focus on neurodegenerative diseases and the collaborative efforts within the scientific community position TELL as a timely and crucial player in addressing the challenges of brain health assessment.

It's About Fitness

TELL's groundbreaking ability to offer a comprehensive, noninvasive, language-based evaluation of cognitive performance is a monumental breakthrough. In our technologically advanced, avant-garde society, the demand for nonintrusive solutions is paramount, particularly when considering unique patient demographics and the need for cost-effective alternatives. TELL not only meets these needs but also distinguishes itself significantly in this arena.

The Heart of TELL and the Tale of Two

Our team at TELL has an unwavering commitment to democratizing access to crucial insights into brain function. Personally, and professionally, we hold the belief that everyone, regardless of their background or resources, should have the ability to comprehend and monitor their brain health in real time.

This value is deeply embedded in TELL's mission to serve as the premier platform for speech-based brain function assessment. At the core of our mission is the conviction that speech analysis will be as commonplace and accessible as routine health checks like measuring blood pressure or checking your temperature.

WhyMeRightNOW.com

This dedication reflects not only a professional commitment but also a personal belief in the transformative power of knowledge. By making speech analytics an everyday tool, we aspire to empower individuals to take control of their brain health, fostering a world where informed decisions lead to healthier, longer lives. TELL's value is thus intricately tied to the broader vision of democratizing brain health insights for the benefit of all, with TELL becoming the go-to place for checking brain performance and analyzing speech.

> **Democratizing brain health insights**

TELL in 2028: Rewriting the Rules

Envisioning the future, the 2028 incarnation of TELL presents a compelling image. At this time, we have successfully penetrated the US market, making a significant impact in Tampa and extending our reach across Florida's Latin and non-Latin +55 population.

By this time, our expansion is not merely national, but we have taken the global market by storm. Our team has doubled in size, and our influence is now dispersed across the worldwide markets.

We have welcomed an exceptional COO to our ranks, whose expertise will guide us in scaling our operations effectively. We are also deeply engaged in strategic partnerships with universities and pharmaceutical laboratories, with a shared goal of halting neurodegeneration. In the 2028 landscape, TELL is not just a participant, but a game changer redefining the rules of the industry.

What makes TELL stand out in the marketplace is our dedication to redefining how we investigate neurodegenerative illnesses through modern-day technology.

Our innovation lies not only in the accuracy of our AI-assisted speech analysis but also in our versatility of use. TELL's area of expertise extends to its adaptability across various linguistic and cultural contexts as we've made it a priority to offer assessments in more than one language. Our aim is to cater to an international audience. This dedication to addressing access inequalities in various populations is entirely unique.

Shared goal of halting neuro-degeneration

Our continuous research endeavors underscore our unwavering commitment to medical precision and consistent enhancement. One of TELL's important objectives is to surpass cultural barriers in order to propel neurological research forward and offer accessible, tailored brain health evaluations. By promoting an inclusive and collaborative environment, our goal is to transform perspectives and contribute to a healthcare approach that is more representative and precise.

A healthcare approach that is more representative

WhyMeRightNOW.com

Embracing the Rich Diversity

TELL stands for inclusivity, recognizing and embracing the rich diversity of linguistic and cultural variations. Our technology is made to pair with anyone, making sure it connects with risk populations from various backgrounds. Our collaborations, particularly our involvement inside the ReD-Lat consortium, emphasize breaking down geographical and cultural limitations in medical studies. By actively running with establishments throughout Latin America and the US, we aim to contribute to a more truthful representation of different languages and diverse populations in neurodegenerative ailment research.

Notes from a Founder

As the founder of TELL, Fernando Johann, hailing from Uruguay, is on a journey with TELL to reshape perceptions. As his team navigates the twists and turns of TELL's adventure, from his roots in academia to the worldwide aspirations of rewriting the rules in neurodegenerative evaluation, there's a harmonious combination of commitment, resilience and cultural richness.

TELL's Journey and Cultural Circle

Fernando asserts that TELL's most cherished part of their Latin identity is undoubtedly the distinctive method of connection.

It's a cultural nuance that's challenging to articulate to those outside the circle, but in the Latin community, we embrace each encounter with a special warmth.

The tradition of kissing hello and hugging goodbye is ingrained in our interactions, transcending the boundaries of familiarity. Whether you've known someone for years or are meeting them for the first time, this gesture is a universal expression of closeness.

The pandemic did cast a temporary shadow on this cherished practice, disrupting our customary greetings. However, what sets us apart is our resilience. Despite all the challenges, our cultural embrace quickly rebounded, a testament to the enduring spirit of our souls. It's a unique aspect of being Latin that not only survived the global upheaval but emerged right back again, underscoring the depth of our connections in a way that sets us apart from much of the world.

> What sets us apart is our resilience

Tampa Bay Wave: A Fortuitous Discovery

Our association with the LatinTech Accelerator program at Tampa Bay Wave has proved to be a fortuitous opportunity for us at TELL. We had been seeking avenues to enhance our business prospects when we serendipitously discovered Tampa Bay Wave. This venture commenced as we participated in a virtual event, with the objective of acquiring insights into investment strategies.

WhyMeRightNOW.com

The host brought up the organization and mission of Tampa Bay Wave, and as the application deadline approached—it was the next day to be exact—we plunged headfirst into the process. We've never been so filled with gratitude and excitement being a part of this cohort group.

This journey has certainly been more than a mere chance encounter. It was a compelling urge that propelled us, a readiness to seize an opportunity, to explore uncharted territories and to initiate our seed round. This was a strategic move, a decision influenced by a blend of calculated risk-taking, preparedness and the invaluable guidance of our supporters. In essence, our "why" is a calculated advancement, a choice made possible by a stroke of luck and a wealth of sound advice. For all this and much more we are more than enthusiastic and absolutely excited for what lies ahead!

> Our "why" is a calculated advancement

TELL provides an AI-driven speech analysis technology for comprehensive brain health assessments, offering scalable, personalized, reliable, and accessible insights to empower individuals, healthcare experts, and pharmaceutical research for enhanced brain health monitoring and optimization.

TELL

Montevideo Department, Uruguay
Founders: Fernando Johann & Adolfo Garcia
tellapp.org

CHAPTER 2

A Proud Badge of Recognition

> "Failure isn't a roadblock but a gateway to new opportunities."
>
> —Ismael Villanueva,
> Co-Founder and CEO, Miss Berni

Education is not a simple, linear endeavor. The diverse skills within our team at Miss Berni enable us to tackle the complexities of the teacher shortage crisis in the US, from a well-rounded and expert viewpoint.

This is the recognition that addressing these issues necessitates a broad-based strategy that transcends the conventional landscape of the education sector. Our team provides a comprehensive approach that reaches beyond traditional boundaries and the norms established over the last two decades.

Unwavering alignment with the educators

Miss Berni is not just a company that bridges the gap between Latin American teachers and schools in the United States. We view ourselves as a fresh, dynamic force in the realm of education and teaching. Our unique selling point lies in our unwavering alignment with the educators we serve. We are involved in every aspect of their journey, from professional growth to recruitment to the pursuit of new opportunities. Miss Berni is always present.

We have plans to establish our presence in Latin America, the United States, the Emirates and Asia. Our goal is to become a benchmark not just in the field of education, but also in the realm of impact-driven entrepreneurship. How do we plan to achieve this? Our assurance lies in our ability to implement policies that facilitate a more fluid global mobility of educators. By doing this, we believe that the collaborative efforts of teachers worldwide will significantly improve the quality of education on a global scale.

WhyMeRightNOW.com

The Turning Point for Miss Berni

In 2022, we observed a notable increase in the rate of teacher turnover, which was double that of past years. This same year, as we began our journey with Miss Berni this change presented our organization with a special chance to tackle the escalating demands in the field of education.

Our team didn't merely view this as a venture but rather as a call to action. This particular moment highlighted the crucial part Miss Berni could play in guiding us through the changing terrain of education.

As most of us understand, the educational industry faces substantial demand for teachers. This prompted changes in districts and state laws to accommodate. This dangerously high demand allowed for some flexibility in teacher hiring. This change created a positive environment for our team to offer innovative and adaptive solutions to effect positive change. The stage was set so to speak with these changes for Miss Berni to respond not only to immediate challenges but also be in a position to pioneer new methods to shape the future of education.

> Changing terrain of education

It wasn't solely the external environment that drove our company to enhance its service offerings and implement progressive organizational changes. A significant portion of the positive transformation during this period was attributed to a shift in our leadership ethos. Our leadership encompasses a broad spectrum of fields, including professional education, architecture, sustainability, law, business and technology. The leadership mindset for Miss Berni represented a substantial commitment, both personally and professionally, reflecting a profound belief in the potential and influence of our team.

We believe fully in our ability to make a significant difference. With this unwavering dedication, our team sets the stage for a mission fueled by knowledge and a profound passion to contribute meaningfully to the education sector.

Redefining Excellence—the Miss Berni Way

Our approach at Miss Berni goes beyond addressing the teacher scarcity crisis, we aim to redefine excellence in education by introducing an exclusive technique.

First, we believe it's about showcasing the exceptional preparedness of Latin American teachers. We position Latin American educators as valuable individuals capable of creating a significant impact in the United States by highlighting the inherent strengths of those Latin American teachers.

Second, our approach toward cultural variations is not one of the disadvantages but of gains. We see cultural diversity as a leadership position within instructors and schools.

Our commitment is clear: to contribute to enhancing and spreading the incredible talent found in Latin America globally. This perspective acknowledges that talent knows no borders and can thrive in diverse environments.

Talent knows no borders

As you read this chapter you may now see how Miss Berni stands out. We know it is because we embrace our uniqueness. Our attitude is disruptive from the start, challenging norms by connecting teachers across borders to address this shortage crisis. This firsthand experience positions us to understand the needs and challenges on both sides of the issue as we align with them in their schools.

The Motive Embedded

Over 500 applications from across Latin America were received by Tampa Bay Wave for the LatinTech program. Out of this pool, Miss Berni was chosen for this cohort group. This is a testament to the passion and motive embedded in our team, showcasing the collective brilliance within this group of tech pioneers, planning to introduce groundbreaking initiatives.

The significance of our selection goes beyond mere profitability. We showed our unwavering dedication to improving education, to being a driving force in addressing the enduring teacher shortage crisis in the United States.

For our team, this recognition of being a part of this Tampa accelerator is not just a feather in our cap. We've used it as a catapult to continue our contribution with our technology in the field of education and supporting teachers. Our ongoing dedication to alleviating the teacher shortage crisis underscores our commitment to a motive beyond our achievement. We took this opportunity now as a time to communicate our message resonating in boardrooms and classrooms where teachers are the unsung heroes.

WhyMeRightNOW.com

Miss Berni envisions playing a more substantial role in the education landscape, fostering the best of AI innovation and excellence. The journey doesn't stop at being chosen. It evolves into our continued pursuit of addressing educational challenges and making a difference in the lives of teachers and students.

> **Playing a more substantial role**

At the heart of Tampa Bay Wave, beyond the invaluable practical guidance and strategic wisdom shared, what truly shines is the team's unshakable faith in Miss Berni's vision. In the dynamic world of entrepreneurship, where doubts may occasionally creep in, having a titan like the Wave standing firmly behind us, the founders and our company is a quiet yet potent source of motivation. It serves as a constant reminder that our ambitions are not mere dreams but attainable milestones. This unwavering support from the Wave is a testament to their belief in our potential and their commitment to our success.

> **Potent source of motivation**

Embracing Failure and Paving the Path Forward

For our teams' viewpoint, failure isn't a roadblock but a gateway to new opportunities. This ideology is communicated by our founder as he reflects on both his professional and personal journey. The first of his two failures was his involvement in a 3D visualization agency in the architecture and construction industry. The mistake he admits is not committing wholeheartedly and fully. This experience taught him the importance of unwavering determination.

The second failure happened during his tenure at a prestigious architecture firm on Wall Street. When faced with burnout and a desire for personal and professional change, decisions were made based on immaturity in his professional life. These pivotal decisions led to a profound realization—life is more of a marathon than a sprint.

Out of this awareness, Miss Berni's organizational culture was born, a company focused on making big changes in education, with the insight and patience of being unwavering in the long-term objectives of the team.

Navigating Challenges with Resilience and Authenticity

In our unpredictable entrepreneurial journey, having a Plan B is not a signal of doubt but a testimony to resilience. For Miss Berni, Plan B is not an exit strategy, it's a commitment to allocate additional resources to ensure the fulfillment of Plan A.

> Plan B is not an exit strategy

WhyMeRightNOW.com

Recognizing the strength of a startup lies in its potential to pivot. We embrace the agility that allows us to adapt and refine our techniques, ensuring that Plan A not only works but flourishes. The dedication to vision is a team sport, and our company team supports fully our Plan A.

Capabilities Are Amplified Exponentially

As a Latin America company founded in Chili, outside our business endeavors, we love the incredible food produced in our community. The richness of flavors and the communal experience of sharing food embody the essence of Latin culture. It's a reminder that even amid bold dreams and transformative projects, simple pleasures and cultural connections play a significant role in the journey of all our lives.

As we wrote earlier in this chapter, we wholeheartedly believe that our Latin educators are valuable, uniquely gifted instructors. Our conviction is unshakable that our company's capabilities are amplified exponentially due to our rich Latin heritage and distinctive global perspective. This unique blend of culture, technology and our worldview makes us stand out, empowering us to achieve at even greater heights.

> **Our conviction is unshakable**

Miss Berni is a "Teacher Hunter" who recruits Latin American teachers for schools in the United States.

Miss Berni

Santiago, Chile

Founder: Ismael Villanueva

missberni.com

CHAPTER 3

Pioneering Everyday Health Monitoring with Wearables

> " We are experts in sensors, wearables and physiology and we want you to be the CEO of your own health and wellness. "
>
> —Christy Fernandez-Cull,
> Founder and CEO, Davinci Wearables

The COVID-19 pandemic underscored the importance of health. In its aftermath, a significant transformation occurred in the field of wearable fitness trackers and overall well-being. The rise in the elderly population, combined with a heightened emphasis on fostering a resilient and healthy lifestyle, has ignited a significant upsurge in the enthusiasm surrounding wearable health and fitness technology. Devices such as smartwatches and smart home gadgets are now being sought after not only for fitness purposes but also for enhancing everyday health awareness.

Wearable devices have transformed health monitoring and management, offering real-time data and a motivation for a healthy lifestyle. These smart gadgets are invaluable companions in our journey toward improved well-being. At Davinci Wearables, we recognize this shift and are at the forefront of innovation. We combine miniaturized sensors, continuous health data monitoring, advanced AI algorithms and the need for personalized insights to seamlessly integrate wearables into everyday clothing. Our expertise in these technologies, including clinical data, female physiology, sensors, biomarkers and algorithm design, sets us apart in this field.

Enhancing everyday health awareness

Incorporating sensors into clothing is an innovative strategy that aligns with evolving customer preferences. We view wearables not as mere accessories but as essential components of everyday attire, enhancing both health and performance.

These principles are fundamental to Davinci Wearables' approach, particularly in the precision-focused realm of fitness tracking.

Empowering Female Athletes with Davinci Wearables

Davinci Wearables has made a significant impact, especially among female athletes and individuals seeking on- and off-the-field health and wellness solutions. Our journey has been characterized by strategic partnerships.

In an investment landscape that often overlooks black, brown and multi-hyphenated founders, Davinci Wearables distinguishes itself through its dedication to social impact and a software-centric approach. Despite less than 2% of funding typically going to black and brown founders, and even less for multi-hyphenated individuals such as female-Hispanic founders (less than 1%), we are challenging the status quo. Our goal is to have a meaningful influence on how young women engage in conversations about their bodies after puberty.

Davinci Wearables aims to enhance female confidence and awareness across different life stages, from puberty to menopause, achieved through both social and physical means.

> We are challenging the status quo

With our groundbreaking Davinci Champions program and advanced mobile app, we cultivate a vibrant social setting. Our aim is to empower women with detailed daily insights into their bodies, enabling them to leverage this information for optimized training regimens. This knowledge equips them to confidently and effectively manage their daily, weekly and monthly schedules.

Our entire endeavor is focused on empowering active lifestyles by making garments smart and providing actionable insights accessible through our mobile app. The mobile app utilizes AI-powered algorithms to provide personalized recommendations for promoting overall well-being, enhancing performance and reducing mental stress. The app's algorithms can use both the data from Davinci's smart garments as well many other common health-tracking accessories such as smartwatches and smart rings.

Unveiling the Davinci Wearables Team

Our journey with our team is more than a business venture. It's a collective effort of passionate innovators, athletes and women who endeavor to leave an indelible mark on the intelligent-clothing landscape.

Innovation defines our path, backed by patents, research papers and the successful execution of products at scale. As a team comprised of former and current athletes, we intimately understand active individuals' unique needs and challenges, particularly the needs of most women.

At our core is our commitment to innovation, and with our diverse and accomplished leadership team our company is set apart. As a woman-led Latinx organization, Davinci Wearables boasts a team of experts ranging from PhDs to MDs who have held leadership roles at Fortune 100 companies like Apple, Novartis and Google. Physiology, sensors and wearables aren't just areas of expertise for us, they are integral to our identity. We're not just leaders, we lead by example. Our dedication to excellence extends beyond the present. It's a thread that runs through our historical achievements and continues to shape our current endeavors, all led by our incredible management.

> We're not just leaders, we lead by example.

Beyond Davinci Wearables—Future Aspirations

Our vision extends into the realms of venture capital and strategic management. A key aspect of our future direction is the development of a venture capital thesis, which will serve as a guiding framework for making impactful investments in emerging technologies.

Within the venture capital arena, our key areas of focus revolve around cutting-edge technologies, with a strong emphasis on AI/ML, data privacy and security. We envision a future where these technologies seamlessly converge with critical sectors, particularly in mobility, augmented reality and digital health. Our objective is to foster growth and innovation in these essential verticals through strategic partnerships and investments.

The Cultural Fabric

Our mutual dedication to family holds great importance in our journey. *Family* is not merely a word but the foundation of our team's principles and deeds. It symbolizes a bond grounded in trust, love and the liberty to express and feel a spectrum of emotions. These values are an integral part of our cultural identity and reach beyond our immediate circles to embrace the wider community we interact with.

> **Dedication to family**

Collaborations Are Key

Being part of the inaugural LatinTech cohort at Tampa Bay Wave fills me, Christy Fernandez-Cull, with excitement. As the founder and CEO of Davinci Wearables, I see great potential for collaborations, pilot programs and mentorship opportunities through this partnership that aligns with our company's goals. This collaboration serves as a crucial stepping stone as Davinci envisions a transition toward continuous health monitoring to enhance women's health and athletic performance through preventive measures.

We are deeply thankful for the vibrant collaboration we've experienced with Tampa Bay Wave, which has brought unprecedented attention to Latin tech entrepreneurs on a national scale. This emphasis has not only met but exceeded our expectations, highlighting the tremendous potential and significance of such partnerships.

Our team is not just hopeful but filled with unwavering certainty and determination in the direction we are already forging. Our product goes beyond clothing; it stands as a concrete testament to our dedication to empower women, inspiring them to proactively improve their health throughout their hormonal inflection points and overall wellness journey. Our relentless commitment to understanding and enhancing women's unique health experiences drives our mission to bring about transformative change in our specialized market.

A wearable health and wellness platform for active females, offering a proprietary liner, a user-friendly app that syncs data with 3rd-party tools, and tailored sport-specific recommendations for peak performance, preferences, providing safe options.

Davinci Wearables

Sunnyvale, CA, USA
Founder: Christy Fernandez-Cull
davinciwearables.com

CHAPTER 4

Revolutionizing Recruitment

> Time and objectivity are vital, and that is why both are part of our core objective.
>
> —Daniel Pardo, Co-Founder, Hitch

The global corporate landscape is riddled with the costly conundrum of mismatched hires, draining billions from companies' coffers each year. Enter Hitch, a trailblazing SaaS (Software as a Service) company, poised to revolutionize the recruitment process. Harnessing the power of AI and machine learning, Hitch supercharges the hiring process, transforming it from a potential pitfall into a powerhouse of precision and efficiency. This innovative SaaS is not just a tool but a game changer, set to redefine the effectiveness of recruitment processes in and outside Latin America.

Accepting the knowledge that human capital is crucial to all enterprises, Hitch considers this their compass as they seek to dominate the human resource sector.

It's obvious that by accelerating the hiring process, companies will receive higher-quality employees faster. Instead of ravaging through files for weeks, comparing the different achievements of different candidates, Hitch's software can determine the right candidate within a fraction of the time it takes to make the right hiring decision.

> **Human capital is crucial to all enterprises.**

Hitching Success to Goals and Characteristics

Landing in the software development industry in Mexico City, co-founder Daniel Pardo saw the dire need for automation within companies for recruitment. Noticing a surge in this industry as current practices were losing efficiency for their clients. While implementing artificial intelligence and machine learning into their practices, Pardo's team found the solution. The solution was Hitch, founded in 2020.

Using more than comparing the skills of the candidate, the data-driven tool Hitch developed can select the perfect candidate for companies. Hitch is all about "connecting the dots in the right way." They find the data that recruiters often overlook.

Hitch is set apart from competitors because it encases the entire recruitment process, creating work for nearly every position available and focusing on SMEs (small and midsized enterprises). This laser focus discovered an underserved niche of customers in need of their services.

> They find the data that recruiters often overlook.

The team at Hitch primarily works in four sectors: business development, engineering, information technology and human resources. Those segments make Hitch's operations possible as their charter is executing unique strategic approaches to solving issues within these industries. Hitch's teams are consistently working toward refining the process for recruitment.

Guiding Growth—Leadership Insights in Practice

Hitch's team coalesced at the perfect moment, just when the industry needed them the most. They are firm believers in the power of unity, the strength that comes from overcoming hurdles together and the triumph of shared objectives. In their dynamic workspace, *synergy* isn't just a buzzword—it's the driving force. They understand that a company thrives when each individual is at their peak performance. This principle of leadership is something that they never let fade into the background.

> Each individual is at their peak performance.

Hitch is poised for expansion into the US Latino markets, a move that promises to fuel its growth. As a trailblazer in this arena, Hitch's potential for success is amplified by the right mentorship and a deep understanding of digital marketing. This knowledge will be instrumental in creating networking opportunities and enhancing overall exposure, paving the way for this ambitious goal to become a reality.

No doubt, Hitch is on the fast track to becoming the global leader in software solutions, transcending boundaries and making a mark not just in the US but across the world.

Embracing Values and Clarity

One of the team's most significant values is authenticity. Authenticity is about being transparent with who you are and translating those behaviors into clarity in communication within your leadership and team. Transparency doesn't necessarily extend to completely disclosing all your information, especially information that might be personal. It is more of a welcoming invitation to employees and people you come across. It is the mission that your organization is led with honesty and professionalism in all ways.

> **Welcoming invitation to employees**

Additionally, leadership can effectively communicate and guide their team while also sharing some of their weaknesses if the message will improve the overall team's performance. An authentic leader provides knowledge for the team that benefits not only the company but also its employees. It is essential to remain honest as a leader, as your team will operate effectively when their leader maintains transparency.

Aside from its values, Hitch has long-term objectives for the brand. Within five years, Hitch aims to evolve into a one-stop store where all company essentials can be purchased under their brand.

> **Long-term objectives for the brand**

While they are currently heavily focused on the business development and human resources sectors, they want to

WhyMeRightNOW.com

extend their reach outside this area. This means broadening their base, and including various components of human resources for new vertical markets.

These components will be implemented within the human resources sector to ensure employees gain valuable insight into necessary skills for success while cultivating an environment that attracts better-suited candidates. This expansion will not only be accessible by these companies, but implementing the components will foster a positive, collaborative working environment across different industries.

Embracing Latin Culture's Vibrancy

As companies like Hitch make progress, ideologies exist based on bias. These beliefs are beginning to dissipate due to positive minority representation. In breaking perceptions based on cultural background, Daniel Pardo is poised to succeed as a Latino founder of an extremely successful startup company. This is in hopes of encouraging people from similar and diverse backgrounds to recognize the limitless potential in the international business arena.

As Hitch expands, many individuals from younger Latin generations will certainly see the power and success from looking at a Latino founder and witnessing the achievements of an all-Latino business team.

All it takes sometimes is to see someone who looks like you in a position you want to be. This can be a powerful example of the achievement you may desire in your own career. Hitch's presence alone may cause future generations to become even more inspired to become powerful leaders as they see what can be accomplished through Hitch's example.

In addition, Hitch also advances diversity while displaying the strength of being a positive representation of the Latin community.

Clearly, Hitch takes pride in their Latin roots. One of their favorite parts of being Latin is the warm hospitality and upbeat dance beats that produce welcoming experiences in Latin culture. The value for collaboration and workplace synergy is derived from that heritage. Hitch has strategically linked aspects of the Latin culture to their business.

> Strength of being a positive representation

In wrapping up this chapter, it is profoundly impactful to note that Hitch's quest to secure a robust network of essential team members worldwide is set to revolutionize the companies they help and the industries they will expand into in the future.

The groundbreaking structure of Hitch's technology is poised to dramatically enhance the process of securing the ideal candidate for their clients, creating a monumental impact. This is not just a conclusion, but it is a thrilling anticipation of what's to come.

An end-to-end HR SaaS utilizing advanced technology for autonomous recruitment, offering a self-service platform with intelligent video interviews and gamified cognitive skill assessments without human contact.

Hitch

Mexico City, CDMX, Mexico

Founders: Daniel Pardo Salazar and José Miguel Arreola

hello.gethitch.ai

CHAPTER 5

The Potential of Latino Talent in Tech

> "Life is not a dress rehearsal. Love the process."
>
> —Nick Lopez, Founder, Prosal

We find ourselves at a pivotal moment in the economy. Starting a business has never been more accessible, and at Prosal, we firmly believe that Florida is the epicenter of South American tech innovation. Our team is proud to be a part of this exciting movement.

Our vision is to redefine the landscape of RFPs (requests for proposals), traditionally perceived as equitable and transparent, by acknowledging the inherent bias that disadvantages smaller agencies in competition with larger, more resource-rich businesses.

We recognized that merely leveling the playing field was insufficient. We needed to revolutionize it.

We see RFPs as the gateway to scaling a business on demand. Additionally, the rise of remote work has simplified the process of providing professional services from afar. A larger talent pool and a wider range of options can be used to result in improved outcomes for our client responses to proposals. We hope to maximize the immense talent pool in this global network.

Prosal's Unique Approach to Product and Advocacy

The differences between what we provide compared to our competitors go beyond just a functional product. We place a high emphasis on thoughtful user design, ensuring that our products are not only effective but also user-friendly and intuitive. We prioritize user experience.

This experiential design is crafted with a deep understanding of our customers, making our platform feel more like social media than a central authority RFP aggregator. This intentional consumer layout prompts our customers to recommend our team and refer potential clients to us.

> More like social media

But it's not just about the excellent user experience. We stand out because we provide the only platform that advocates on behalf of the service provider. Unlike most RFP-associated platforms, which are constructed from the angle of the service seeker, we flip the script. Our belief is that RFPs are or should be two-way streets, and high-quality engagements manifest when each side comes with their best on a playing field that suits all teams.

> We flip the script

This unique approach and one-of-a-kind methodology underscore our dedication to developing a platform that caters to small and medium-sized businesses with the same level of service and functionality that larger enterprises have come to expect. Our emphasis on intentional layout and advocacy will stay at the center of Prosal's identity.

WhyMeRightNOW.com

Empathy, Efficiency and Unveiling Hidden Talents

Our founder, Nick Lopez, places a premium on empathy and performance, two values that shape his method of leadership at Prosal. Lopez believes that when you understand your coworkers and employees on an emotional level and relate with their plight, it's much easier to find someone's "all-star" qualities and hidden talents. This perspective allows our management to identify where an individual's *zone of genius* is.

This profound expertise allows him to pick out and nurture sectors of genius in individuals, unlocking their total ability for the good of our team and their professional development.

Efficiency is another cornerstone of Lopez's management philosophy. Driven by a preference for beauty in answers, he seeks the fastest and best routes from A to B. His background in mechanical engineering has equipped him with the ability to examine and execute speedily the sought-after solution.

His proficiency is getting to the core of issues or problems, homing in on them and constructing solutions using available resources to maximize efficiency for our company team.

All this is thriving in the face of complicated and esoteric challenges in the world of professional services today. It becomes no longer just an activity; it is a professional assignment, an unstoppable force to solve problems that others may deem unsolvable. This leadership quality in not only keeping a calm mindset but in problem-solving to lead our team through the thicket has catapulted our company to greatness in many ways.

Cultural Richness, Fertile Environment

Prosal is an organization that stands as a testament to the undiscovered talent in the Latino network, specifically in the tech world. Our charter assignment revolves around empowering Latino-owned organizations to not only compete but to triumph in lucrative business opportunities that exist in requests for proposals.

RFPs predominantly originate from the government, and with our specialized comprehension of such proposal requirements, we establish unique and empowering partnerships for our clients. We hold the belief that Florida, with its abundant cultural richness and technological diversity, provides an exceptionally fertile environment for achieving victories in tech innovation within the Latin community.

> Victories in tech innovation

By unlocking the capability of Latino expertise in tech, we will be a part of reshaping the narrative of what is feasible inside the global expertise of services in and outside Latin America.

The Unyielding Spirit and Future Aspirations

Our spirit at Prosal embodies that of staying power. We view challenges not as failures but as research possibilities. We adopt this unique perspective. We steadfastly affirm that the concept of failure is nonexistent. Failure is simply a novel insight that provides a new learning opportunity.

> **Failure is nonexistent**

Our founder is fond of saying that he believes the one who is having the most fun, *wins*. Our team echoes this theology as we firmly believe that the person who enjoys their work the most, often comes out on top. As we strive to create the most advanced and efficient technology for our clients, we also wholeheartedly embrace an attitude of relentless enthusiasm.

Breaking Stereotypes and Building Something Big

As we are a US enterprise, perceptions based on a cultural and historical past can be uncertain. However, one issue is apparent, the team at Prosal is building something monumental. Our vision extends beyond just commercial enterprise—it is about shattering the stigma that building a vastly worthwhile company and fostering an equitable environment based on fairness will be impactful on many fronts.

The willpower to break stereotypes is important. We imagine a future where our team functions in a world that doesn't set up barriers due to cultural differences. Our experiences have shown us that in the equitable or inequitable realm of RFPs, the victorious bid should be determined by the abilities, expertise and innovative solutions offered in the responses. We imagine and work to implement this environment, where all comes out of a more than equal playing field for all respondents.

> Doesn't set up barriers due to cultural differences

Commitment, Risk and the Singular Focus on Plan A

It's an all-in technique, for in the Prosal sector, it's either a complete commitment or a not-at-all response. When the stakes are as high as they are for success, there is only a Plan A. For Prosal, the idea of building a Plan B isn't an exercise we are interested in spending time on. The time is better spent as extra time on Plan A.

> There is only a Plan A

In fact, the existence of a Plan B can sometimes create an unnecessary safety net, which might lead to a lack of commitment to Plan A. It can subtly imply that failure is an

WhyMeRightNOW.com

option, which might affect the overall motivation and drive to make Plan A not work.

Last, it's important to remember that sticking with Plan A doesn't mean being inflexible. It's about making adjustments and improvements to the original plan based on new information or changes, rather than jumping ship to a whole new plan.

As this chapter closes, it leaves us with insights into the trajectory of Prosal's endeavors in the tech space. With our expertise in this channel, leveraging advancements in AI, we most certainly have begun to level the playing field, making it easier for smaller agencies to compete with large, higher-resourced counterparts. This is a story that weaves together a team of professionals with a rich attitude of success, an appreciation for cultural impact, and a profound know-how that is unmatched in the world of RFPs responses.

Prosal is *the* place for RFPs.

Prosal is a B2B SaaS platform that helps agencies and consultants find and win requests for proposals.

Prosal

Miami, FL, USA
Founders: Nick Lopez and Alfredo Ramirez
prosal.io

CHAPTER 6

Nufi's Important Work

> *We don't just check the boxes, we dig deep, tapping into an extensive range of data sources exclusive to us in Mexico.*
>
> —Hans Villa, Co-Founder and CEO, Nufi

The timing couldn't be better for Nufi! As the business world is rapidly embracing digital transformation, the importance of reliable identity verification is becoming a top priority. This is especially true in the finance and tech sectors. With the LatinX network making its mark in the US marketplace, there's a growing need for services that can facilitate agreements across borders.

Nufi, founded by Hans Villa, excels in offering comprehensive facts about people and businesses from Mexico in a marketplace where no other resource exists. We aim to provide crucial data, fostering acceptance as accurate, mainly for those with ties to Mexico coming into the US market. This adventure excites us as we bridge gaps in this sector, taking advantage of the networks and information provided with the aid of the Tampa Bay Wave LatinTech Accelerator.

> We aim to provide crucial data

Integrity is Nufi's guiding principle, both professionally and personally within our management team and in our marketplace. Being honest, fair and moral is at the center of who Nufi is. We recognize the critical role of providing correct and reliable information to foster trust and provide our clients with the knowledgeable choices they need.

Integrity isn't only a value. It motivates us to be accountable and devoted in each interplay, whether with clients, with partners or within our team. It ensures that our work products and services are grounded in building and maintaining trust.

Nufi is here to change the game and we are doing it by showing what we can do beginning in Latin America. We are not just an agency based in Mexico. We serve as a dynamic hub for innovative ideas and exceptional data production. Our success story is a testament to the fact that effective and straightforward tech solutions can emerge from any location. We are actively challenging traditional notions and simplifying processes for our clients.

Nufi's Vision and Approach

We see Nufi becoming a massive resource to ensure people are who they say they are. In just a few years, by 2028, Nufi will transcend its critical role in Latin America and become a globally renowned powerhouse. Our services will be sought after by businesses worldwide, who will not only require but demand verified, truthful data to authenticate the identities of individuals. Our purpose is crystal clear: to establish ourselves as the unrivaled leaders in delivering top-notch, accurate and verified data to the most esteemed companies, empowering them with invaluable insights into their current and prospective workforce.

> **People are who they say they are**

Let's delve into why Nufi stands out as not just an ordinary player in the sport but as the true MVP. First, our method of verifying a person's identity goes beyond surface-level scrutiny. We delve deep into the intricacies to ensure utmost accuracy. We have many unique resources of facts available only in Mexico, giving us the complete picture. Next, we can extract a single piece of

WhyMeRightNOW.com

information, and from that we can paint a complete picture of a person's identity. It's a thorough system, making this business less complicated for all.

Our company offers a custom, user-friendly platform that seamlessly facilitates instant exams and a range of other services, ensuring a seamless onboarding experience. With our deep expertise, user-centric approach and unparalleled versatility, we stand out as a truly unique solution for our clients.

> An opportunity to test a new hypothesis

At Nufi, we're all about being scientific and methodical in our approach, which means we have plans ranging from A to Z. We operate on the principle of generating hypotheses and rigorously working to validate them. This mindset allows us to focus on execution and adaptability rather than being fixated on expected outcomes. If a plan doesn't work as anticipated, we see it as an opportunity to test a new hypothesis. This approach ensures that we're always ready to pivot and evolve, continually seeking the most effective ways to serve our customers and achieve our goals.

Nufi's Cultural Connection and Community Bond

Something that is incredibly dear to the founders of Nufi is their deep love and affection for the Latin community. This community holds a special place in their hearts, as it embodies a strong sense of togetherness and unity. The founders truly believe in the power of relationships and the importance of fostering genuine connections. This belief is deeply ingrained in our company's values and influences every aspect of who we are. Our cultural roots, which are deeply intertwined with the Latin community, bring an unparalleled passion and drive to everything we do. We are dedicated to achieving success and distinction within our own community and all the cultures we care about in the world.

Being part of the Tampa Bay Wave LatinTech Accelerator helped us understand the network spirit of Tampa. We heard how the enterprise and tech network in Tampa is like a big circle of relatives—open, collaborative and complete with mentorship. This has been our experience as part of this accelerator.

> **Tampa is like a big circle of relatives**

This support became a crucial part of our adventure and even more significant part of our journey. In our time in Tampa, we saw a business environment where everybody seemed to be ready to assist startup enterprises. This friendly atmosphere has driven us to be deeply involved with this community, making our time right now more than about developing our enterprise—it's about being a part of an active and supportive network.

Our business story has not just been about achievement, it is about the significant effect we can make inside the sector we are a part of. The cultural richness and the collaborative business

WhyMeRightNOW.com

environment are like gasoline for our adventure, propelling us forward with a shared sense of reason and purpose.

Limitless Interactions

As we envision our future at Nufi, we quantify it with the significant figure of *one million*. This number represents the countless interactions, choices and premium experiences we've encountered. We assert that our figure may even surpass one million, marking the escalation of our journey as incredibly impactful.

It encapsulates each feedback acted upon, each project turned into an opportunity, and the extensive network of connections we've built. *Million* is a testament to our dedication to our nonstop drive for improvement.

Navigating Tampa's Business Ecosystem

In our journey, one piece of advice resonated deeply, embracing the community spirit of Tampa. Tampa's business and tech community is a true gem. It's not just welcoming, it's a place where *collaboration* isn't just a word, it's a way of life. This insight became our cornerstone, guiding our interactions and shaping our approach. We've found a culture of genuine support and a collective desire to see startups like Nufi succeed. The openness and eagerness to share insights have been invaluable, fostering an environment where businesses grow individually and together. Tampa is more than a location. It's a dynamic community that encourages us to engage, contribute and thrive.

The relationships we've built and the collaborations formed are not just professional connections but bonds grounded in shared goals and mutual success. Tampa's community spirit has encouraged us to not only focus on Nufi's growth but also actively contribute to the growth of others. In this vibrant ecosystem, success isn't a solo journey, it's a collective effort. With its wealth of insights and collaborative spirit, Tampa is a fertile ground for our continuous learning.

> Tampa is a fertile ground for our continuous learning

A Cultural Commitment

At Nufi, our commitment to build trust extends beyond our services, as it is deeply rooted in our cultural values. From our Latin background, trust is a business requirement and a way of life.

Our work reflects this cultural commitment, emphasizing the importance of genuine connections and transparency. In an industry often seen as transactional, our approach is personal. We prioritize relationships over transactions, ensuring that each interaction is not just a business deal but also a step toward establishing lasting trust. This cultural ethos shapes our every decision, fostering an environment where trust isn't just earned, it's cultivated, nurtured and the cornerstone of our identity.

Continuous Learning and Evolving

As we progress through the Wave program and engage with the Tampa community, a fundamental principle guides us—the heartbeat of progress is in continuous learning and evolving. Every interaction and every piece of advice becomes a stepping stone in our journey toward excellence. Learning is not a one-time event; it's an ongoing process, a philosophy embedded in Nufi's approach. It's about acquiring knowledge, about adapting and evolving our strategies based on the wisdom gained from our experiences, and then our opportunity to provide better services to our clients.

Embracing the spirit of evolution, Nufi thrives on this. Every challenge becomes an opportunity to refine our methods, every success is a testament to our ability to adapt. As we immerse ourselves in our quest for excellence, we recognize that progress is a journey of perpetual learning. In this dynamic environment, Nufi is an active contributor to the rhythm of progress, driven by a commitment to constant improvement and innovation for all our current and future clients.

Nufi streamlines Mexico's background checks, offering swift, affordable identity verification for businesses, enhancing trust and compliance with an AI tech-driven approach.

Nufi

Monterrey, Nuevo Leon, Mexico
Founders: Hans Villa and Ilich Nunez
nufi.mx

CHAPTER 7

Global Growth and Learning from Experience

> " In business, it's now not just about what you have but how you operate with what you've been given. "
>
> —Javier Cuello, Co-Founder and CEO, H+Trace

From my formative years, the virtues of humility and perseverance were deeply ingrained in me. These principles have since become the bedrock of my approach to both my personal life and our business. The significance of humility is particularly highlighted when our clients express satisfaction upon the fulfillment of what we said we would do for them.

Our company, supplies meticulously accurate data, pivotal for optimizing supply chain operations within organizations. Our expertise is particularly focused on eradicating or significantly reducing laboratory errors for our healthcare clients. We offer data-driven, innovative solutions designed to create processes that yield unparalleled levels of efficiency for our clients.

> **Data-driven, innovative solutions**

While we launched our services in Argentina for commercial validation, we then expanded with great success to Latin America. We now have the momentum and possess the drive to firmly position ourselves in the developed world with a sophisticated solution that will yield tangible outcomes for all targeted client communities.

Leveraging Unique Perspectives for Market Leadership

At H+Trace, our outlook is uniquely shaped by the specialized knowledge required in unconventional markets. Our experience in operating within business environments in developing nations, which present additional challenges, has provided us with a valuable learning perspective that aids in refining our technical business offerings.

This perspective goes beyond the conventional, presenting insights and capabilities that may not be present in privileged environments.

The hurdles we encountered in these less conventional markets were not limitations. We saw them as possibilities for growth. Navigating boundaries fostered the competencies inside our work products, where we learned better strategies to be imaginative and adaptive. These characteristics set us apart to provide for an unfulfilled void inside this already highly aggressive global tech environment. In business, it's now not just about what you have but also how you operate what you've been given.

> **Provide for an unfulfilled void**

Not Just a Buzzword

In our adventure, *authenticity* isn't just a buzzword, it is also a guiding precept. It's the popularity that even amid demanding situations, every expression is an opportunity to recognize and come together for a solution in complete disclosure.

By fostering an environment wherein authenticity is widely known, we create a space wherein misunderstandings are minimized and client relationships can flourish. In the

WhyMeRightNOW.com

tapestry of our interactions, authenticity is the thread that weaves lasting connections, reworking expressive boundaries right into a basis for agreement and collaboration.

> Authenticity is the thread that weaves lasting connections

In healthcare, in finance and in all vertical markets there is only one way to come to new, more efficient solutions, and this is with complete clarity, understanding and, of course, authenticity.

Unleashing Potential by Thinking Bigger

For many Latinos, their mindset might revolve around survival rather than global expansion. At H+Trace, we believe in breaking loose from intellectual constraints and self-imposed limits. We believe it's about wondering bigger and imagining outside current surroundings or past biases. It is about reworking our mindsets beyond our experiences.

The shift from survival-targeted wondering to an international and unlimited mindset is like unlocking a powerful force inside us. As Latinos, we supply a wealthy tapestry of experiences and resilience that positions us uniquely as contenders on the global stage. By breaking loose from mental constraints, we can harness our capability to

> Survival rather than global expansion

WhyMeRightNOW.com

live to tell the tale of success on a bigger scale than imagined. At H+Trace, thinking bigger is not just a strategy but the attitude that propels us ahead.

Constantly turning beyond expectations, even when facing cultural prejudices, establishes a unique and lasting reference to humans. Beyond demanding situations, significant bonds created during the client engagement process can last a lifetime, even after the business has commenced. It's a reminder that humility, perseverance and the commitment to exceed expectations for your clients will contribute to commercial enterprise fulfillment. This can and will have a lasting impact on the lives of those we serve and our team.

The Power of Authenticity in Relationships

Our collaboration and partnership with the Tampa Bay Wave team was an absolute privilege and a pivotal stepping stone. Their exceptional expertise in recognizing and uplifting underrepresented talent, such as ours, made our entry into the world's most significant economy possible and truly exciting.

As we had the privilege of being part of the exclusive LatinTechX cohorts, we were introduced to a plethora of unique offerings

> Unparalleled level of expert guidance

tailored specifically to our team and the current stage of our company. The Tampa Bay Wave program, with its vast resources, including an impressive network of community

WhyMeRightNOW.com

and national relationships, left a significant impression on us. However, what truly set Tampa Bay Wave apart was the unparalleled level of expert guidance they provided, helping us confidently navigate through uncharted waters. Their support has been instrumental in our journey, and we highly recommend the program to other startups seeking to make their mark.

The Essence of Responsibility and Preparedness

In business, being accountable is like having a mystery weapon. Usually it's about having a Plan B. For our team, Plan B is more than a backup, it's a hidden approach within H+Trace, equipped to be deployed when needed. The essence of preparedness is always having a Plan B.

It's not about looking ahead to the worst. It's about being organized for unexpected twists and turns in business. It's looking ahead to challenges, whether expected or unexpected, and then holding a contingency plan ready to launch at any time. We believe in the words of Alexander Graham Bell: *"Before anything else, preparation is the key to success."*

> Holding a contingency plan ready to launch at any time

The preparation on our side leverages our unique technology to deliver our exceptional and innovative solutions. Finally, at H+Trace, we understand the enterprise panorama's unpredictability, which is why having a backup plan has been ingrained in our operational philosophy from Day One.

Embracing Global Aspirations

Today, we stand poised to conquer new territories. The journey from survival to international expansion is a testament to our perception that, as Latinos, our capability is limitless. As this chapter concludes, the flame of our passion continues to burn brightly, guiding us toward horizons wherein achievement is aware of no borders.

> As Latinos, our capability is limitless

WhyMeRightNOW.com

AI platform enhancing healthcare and food and beverage supply chains, delivering high-quality data for actionable logistics insights and decision-making improvements.

H+Trace

Miami, FL, USA
Founder: Javier Cuello
h-trace.com

Words Change Bias

> All of you in this room are changing the cultural bias.

—Dr. Richard Munassi, Managing Director, Accelerator Program, Tampa Bay Wave

Quote from November 15, 2023
Global Entrepreneurship Week
Accenture St. Petersburg

What's Your Number Epilogue

Liza Marie Garcia, COO NOW Publishing & Promotion

> Victory belongs to the most perseverring.
>
> —Napoleon

I used to hate it when I heard the term *serial entrepreneur*, being first introduced to it ten years ago when I relocated my family to Tampa from Seattle.

I very incorrectly thought it described someone who had to start multiple businesses because none of them were successful. I couldn't shake the idea that if you were a "real entrepreneur" you only needed **ONE** business to be successful. My misperception was based on my experience as a former IBM software engineer turned vendor when IBM "offered" me a large national account to manage. This began my journey as a tech CEO, bringing my first company to market at year 16 after years of seven-digit revenues. This was most certainly my own perception of what serial entrepreneurship meant, and I was wrong.

Two companies later, I sit as the Chief Operating Officer of NOW Publishing, which is nearing its eighth year. It's my third startup, and I'm looking to start my fourth. Now I get it. I am also a serial entrepreneur, and with this, my number is **FOUR**, but we will get back to this at the end of this chapter.

What's Your Number?

What is the number that didn't stop you from moving in the positive direction of achieving your goal? A personal or professional goal, or perhaps a lifelong dream, that only you understand about yourself and that you know is your destiny. Do you have a number that you moved beyond to continue to try no matter what might have happened? The number of times you went bankrupt to fund your company? The number of times you reorganized your team in attempts to figure out the solution? What did moving through these "tries" take from you? What was that "try again" number that finally got you over your mountain? A mountain no one else but you were climbing.

An Astronaut and a Swimmer

As we collectively put this book together this past year, a movie came out about a Mexican American named José M. Hernandez. His number was **TWELVE**, because it took him that many applications to become a NASA flight engineer. The movie is called *A Million Miles Away* and chronicles José's perseverance and the Latin culture that shaped him. The process for applying to the NASA program is very detailed, and an individual is allowed to apply once a year. Could you wait for 12 years to achieve your dream? He did.

Another "hero" of tenacity is an American named Diana Nyad. Her number is **FIVE**. In 2013, at age 64, she finally achieved her goal to swim the open waters from Cuba to Key West, Florida. Diana had attempted four unsuccessful and quite brutal attempts to swim this route and wrote about how every single person in her network, family and professional life tried to persuade her to give up. She would not stop at four failed attempts.

We Asked Our Latin Tech Founders

As part of the content gathering with the co-authors of this book, we wondered what their number might be. You might be surprised to read what they told us!

WhyMeRightNOW.com

Daniel Pardo
Hitch, Co-Founder

My number is **THREE**. This provides a good balance between tenacity and flexibility, enabling insightful learning from the first attempts.

Fernando Johann
TELL, Co-Founder

My number's a solid **TWO**, first time hitting the college admission test, well, let's say it didn't end in a victory lap. Made a promise right then that the next stumble wouldn't catch me off guard. Been keeping that vow ever since—learning the ropes, the hard way. Life's got a knack for doubling up on things, when it comes to business.

Hans Villa
Nufi, Co-Founder

My number is large. It's **one million**. It's about the countless interactions, decisions and fine-tunings we've gone through as a company. This number isn't just about quantity, it's about the depth and breadth of experiences we've accumulated in our mission. It reflects every piece of feedback we've acted on, every challenge we've turned into an opportunity, and the extensive network of connections we've built.

Ismael Villanueva
Miss Berni, Founder

My number is **two** as I believe failure goes hand in hand with a new opportunity. I have experienced two significant failures. I learned from this that life is more than a sprint, it is a marathon.

2

Javier Cuello
H+Trace, Founder

MY number is **EIGHT**. For seven years, my partner and I tried to get pregnant, and we only succeeded in the eighth year. I don't give up easily, and luckily neither does she. Today, we are happy parents of two children.

Christy Fernandez-Cull
Davinci Wearables, Founder

I have failed so many times at this point in my life that a single number does not come to mind. I've learned through successes but even more through my failures.

Nick Lopez
Prosal, Co-Founder

My number is **infinite**. I've never stopped something because I've failed because I don't see failure. I only recognize learning. This perspective shift allows me to enjoy the journey and see every attempt as part of the process.

It is said that *it's not how many times you fail, it's how many times you start again*. Being a serial entrepreneur means you start building an enterprise again, over and over. You use this "repeat" cycle to learn from your left turns. You then take your belief, fortitude, perseverance and tenacity, your improved skills and your upgraded mindset, to set you apart in a successful way for the next time you try again.

I've now started my fourth company. My family of four has enjoyed living in sunny Florida. Above all, I am proud to be a fourth-generation Latin American. Yes, my number is **FOUR**.

Here are some more enlightening numbers, connected to familiar names.

400
the companies Richard Branson
founded before he hit it big

9
the number of women in Ruth
Bader Ginsburg's law school class

1,500
the times Sylvester Stallone tried
to sell his movie *Rocky*

WhyMeRightNOW.com

5,126
unsuccessful Dyson vacuum prototypes

12

publishers who turned down JK Rowling's *Harry Potter* manuscript

26

times Michael Jordan was trusted to take the game-winning shot and missed

It is imperative to acknowledge the immense and audacious potential that resides within each us. The understanding that success might be just around the corner is a powerful motivator. Each individual's professional journey is distinct, and the exact measure or number of tries it takes to unlock our full potential may remain elusive until it is actualized.

The future belongs to those visionaries who not only dream but also have the determination to turn those dreams into the exact number needed to achieve success!

Do You Want to Make an Impact?

NOW Publishing will help you build your book and deliver your message in a powerful, impactful way.

Everyone has a story to tell and NOW Publishing is here to help them bring those stories to life. Whether you have already written a book and need a marketing partner to promote your story, or have an idea for a book that can change lives and inspire others, we are here to help you turn that into something memorable and marketable.

EMAIL US!
publish@nowscpress.com

Ask about our
90-Day Idea-to-Author Program!

VISIT US!
www.PublishWithNOW.com

In the dynamic heart of the Tampa Bay area, the Tampa Bay Latin Chamber of Commerce stands as a pivotal force for empowerment and innovation within the Latinx business community.

OUR MISSION IS CLEAR:

To bridge the gap between our region's diverse talents and the opportunities within the business and technological landscapes.

The diversity of Tampa Bay is its strength, yet this richness has not fully translated into equal representation in the spheres of business influence and technological innovation. The Chamber is committed to changing this narrative by fostering an environment where diversity in leadership is not just encouraged but expected.

Cesar R. Hernandez, a Founding Member of our Board of Directors, captures the essence of our mission: "We are fostering a community where Latino businesses are not just surviving but thriving. Through our targeted programs and initiatives, the Chamber is more than a network—it's a catalyst for growth and success for Latino entrepreneurs."

Spanning Pinellas, Hillsborough and Pasco counties, the Chamber operates with the spirit of a traditional commerce hub, enriched with a forward-looking agenda. Our offerings range from entrepreneurial workshops to expansive networking events, all designed to elevate Latin business owners and entrepreneurs.

We invite you to join the Tampa Bay Latin Chamber of Commerce, where your aspirations find support, your heritage is celebrated, and your business dreams are within reach. Together, we are not just building businesses; we are shaping the future of Tampa Bay's economy.

"At the Tampa Bay Latin Chamber of Commerce, we're not just participating in the business world; we're shaping society. Our vision is to make Tampa Bay the epicenter of a Latino renaissance, where innovation meets heritage, and together, we redefine the future."

—Danielle Hernandez, President,
Tampa Bay Latin Chamber of Commerce

You are welcome to join this mission!
Email us at ch@omnipublic.global.

BUILT FOR **FOUNDERS**
FUELED BY **COMMUNITY**

TAMPA BAY WAVE, the #1 Accelerator in Florida, champions innovation by merging world-class, industry-focused programs with unparalleled support for early-stage startups. For over ten years, we've nurtured a dynamic ecosystem, elevating founders without sacrificing their equity. Our efforts have been a catalyst for economic prosperity, impacting both local and global communities.

Bolstered by a robust network of mentors, investors, and partners, and sustained through grants and passionate community backing, we are a beacon of progress and opportunity, not just nurturing startups, but fueling the entire innovation economy in Tampa Bay and beyond. We stand as a testament: built for founders, powered by community.

MISSION

We help entrepreneurs transform innovative ideas into real-world solutions and scalable businesses, fueling important social and economic change in Florida and elsewhere.

VISION

Tampa Bay Wave is recognized for world-class programs and world-class startups that are driving Tampa Bay's national reputation for tech, innovation, and opportunities.

"The LatinTech Accelerator, in partnership with the Tampa Bay Latin Chamber and with vital support from U.S. Representative Kathy Castor and the U.S. SBA, marks a transformative step for tech diversity.

We celebrate the ingenuity of our founders of Latin heritage, whose stories illuminate our mission. This program is more than an initiative; it's a commitment to inclusivity, innovation, and community empowerment in the tech industry."

— *Linda Olson*
Tampa Bay Wave President/CEO

www.ingramcontent.com/pod-product-compliance
Lightning Source LLC
Chambersburg PA
CBHW071044240526
45471CB00014B/579